Hydrogen Safety 2024

Topical Conference at the 2024 AIChE Spring Meeting and 20th Global Congress on Process Safety

New Orleans, Louisiana, USA
24-28 March 2024

ISBN: 979-8-3313-1713-3

Hydrogen Safety 2024

Topical Conference at the 2024 AIChE Spring Meeting and 20th Global Congress on Process Safety

New Orleans, Louisiana, USA
24-28 March 2024

AIChE
The Global Home of Chemical Engineers

Printed from e-media with permission by:

Curran Associates, Inc.
57 Morehouse Lane
Red Hook, NY 12571

Some format issues inherent in the e-media version may also appear in this print version.

Printed with permission by Curran Associates, Inc. (2025)

AIChE
120 Wall Street, FL 23
New York, NY 10005-4020

Phone: (800) 242-4363
Fax: (203) 775-5177

www.aiche.org

Additional copies of this publication are available from:

Curran Associates, Inc.
57 Morehouse Lane
Red Hook, NY 12571 USA
Phone: 845-758-0400
Fax: 845-758-2633
Email: curran@proceedings.com
Web: www.proceedings.com

TABLE OF CONTENTS

HYDROGEN SAFETY AND INDUSTRIAL APPLICATIONS AS PART OF THE ENERGY TRANSITION I

HYDROGEN SAFETY AND INDUSTRIAL APPLICATIONS AS PART OF THE ENERGY TRANSITION II

30a Qualitative Hazard Analysis of Bulk Hydrogen Fuel Storage

Alyssa Kuykendall

As the push toward decarbonization and away from oil and gas has strengthened, hydrogen fuel has become increasingly studied and considered as one of the more promising fuel alternatives in this effort. However, before significant reliability on hydrogen fuel can be achieved, and before Hazard and Operability Studies (HAZOP), Layer of Protection Analyses (LOPA), Mechanical Integrity (MI), and Pressure Relief Analyses (PRA) can be accurately assessed, major safety concerns, specifically regarding the bulk storage of hydrogen fuel, need to be addressed and analyzed. The potential hazards of hydrogen fuel are generally understood, but studies of such hazards tend to be quantitative and lack a "cause and effect" analysis, and/or the studies lack inclusion of all possible scenarios. This paper will present a comprehensive, qualitative analysis of hazards associated with storing bulk amounts of hydrogen fuel.

Specifically, this paper will analyze the cause and effect of the following scenarios, why hydrogen poses increased risk for such scenarios, and propositions for potential proactive and/or reactive protections:

- Degradation of storage tank material.
- Explosions and fires resulting from various causes.
- Unexpected vessel heating.
- Joule-Thompson effect during tank filing.
- Unexpected vaporization of liquid hydrogen fuel.

This paper will also discuss proper pressure relief device sizing associated with each scenario as well as pressure relief discharge location.

A qualitative study of all potential hazardous scenarios is crucial in the growing integration of hydrogen fuel. Without a comprehensive understanding of the cause, effect, and progression of a scenario, and the hazardous properties of hydrogen itself, bulk hydrogen storage will maintain significant risks.

The target audience for this paper is anyone developing or using Process Safety Management (PSM) protocol for hydrogen fuel production facilities.

30b Hidden Hazards of Pressure Relief Devices in Hydrogen Systems

Alejandro Jimenez

The energy sector is looking for a sustainable alternative that allows reduce the operational environmental impact to the environment while we can maintain the energy consumption. Specifically, the goal is to find a solution that reduces the carbon footprint emissions.

In this effort, the industry has focused its attention on hydrogen as one of the most promising alternative energy sources using it as an energy production and storage. However, this change brings diverse process safety challenges that need to be considered in this transition. For example, it is known that hydrogen has a higher probability of ignition when released to the atmosphere and the presence of other failure mechanisms as the hydrogen embrittlement has the potential to increase the fire & explosions operational risk.

Given these new hazard characteristics of hydrogen, there is an increase uncertainty related to the risk present when traditional pressure relief devices (PRD) are used in hydrogen systems. Furthermore, although these devices are designed to mitigate a potential rupture due to an overpressure, these could also fail producing leaks acting as the initiated event of an undesired event as a fire or an explosion leading to an uncertainty related to the risk trade-off between the PRD capabilities to mitigate versus the times, they produce an incident.

Due to this uncertainty and motivated by the *Dimond Bar* incident in 2018, a risk informed review was performed in conjunction with the System Risk and Reliability Analysis (SyRRA) Laboratory of Maryland University related to the PRD in hydrogen systems. This presentation will summarize the reviewed performed where hydrogen specific incidents information was collected and evaluated to identify any potential trend. The presentation will present the contributors found during this evaluation showing breakdown on failure modes, operations, and incident types, while comparing the results to the existing knowledge on PRD reliability from the industry.

The presentation will end by providing the most important conclusion on this evaluation and will provide the next step to be performed in conjunction with the SyRRA in order to reduce the uncertainty related to PRD in hydrogen systems.

30c Numerical Simulation of Cryogenic Liquid Hydrogen Tanks with Multilayer Insulation Exposed to Fire

Giordano Emrys Scarponi, Davide Camplese, Robert Eberwein, Aliasghar Hajhariri, Frank Otremba, Valerio Cozzani

In the ongoing energy transition, hydrogen has emerged as a promising alternative energy carrier with a reduced environmental impact. Among the possible solutions to store hydrogen onboard vehicles, cryogenic tanks equipped with multilayer insulation (MLI) appear to be one of the most effective in ensuring high volumetric energy density. MLI systems consist of several layers of low-emissivity material (radiative layers), typically aluminum or aluminum-coated polyester, interleaved with low thermal conductivity spacers to avoid direct contact between the radiative layers. In cryogenic tank applications, these are enclosed within the vessel double-walled shell, working under high-vacuum conditions. Among the insulation systems available nowadays, MLI-based ones have the smallest volume requirements and the lowest weight. Thanks to these features, MLI appears as the preferable choice in sectors where space and weight constraints play a crucial role (e.g., in the transportation sector).

The widespread deployment of LH_2 new technologies poses also challenging questions related to the hazardous properties of hydrogen. The accidental loss of integrity of cryogenic LH_2 tanks might lead to extremely dangerous phenomena, such as Boiling Liquid Expanding Vapour Explosions (BLEVE), Fireball, and Rapid Phase Transition (RPT). One potential scenario that could give rise to this situation is exposure to an external heat source such as a fire triggered by a road accident. Real-scale fire test results suggest that the insulation performance of MLI systems may undergo severe degradation when these are subjected to high temperatures, leaving the tank almost unprotected and leading to failure in a relatively short time.

In this framework, the availability of models able to simulate the tank response to fire exposure is crucial to ensure a safe design and support emergency response planning. Several CFD and lumped models originally developed for pressurized and atmospheric tanks were extended to cryogenic vessels. However, these provide results in line with experimental data only if the insulation system's equivalent thermal conductivity is fine-adjusted to higher values with respect to normal operating conditions. These models do not integrate, in fact, the description of MLI thermal degradation as a result of fire exposure, which was demonstrated to play a crucial role in determining the response of the cryogenic liquid hydrogen tank.

On the other hand, MLI heat transfer models currently available in the literature are suitable for normal operative conditions only and do not address the material behavior under fire exposure.

This work presents an innovative lumped model to simulate the thermal response of MLI-insulated cryogenic hydrogen tanks in fire scenarios. In particular, the proposed approach enables the prediction of MLI loss of insulation performances due to fire-induced thermal degradation, overcoming the limitations of currently available models.

The MLI degradation model is based on the well-established layer-by-layer approach and integrated with sub-models to estimate the material deterioration due to thermal load. Several sub-models were defined to account for different MLI materials. In particular, for polyester-based MLIs, the deterioration of each layer is defined by the apparent kinetic of thermal degradation retrieved from Thermo-Gravimetric-Analysis (TGA) tests. For aluminum-based MLI, each radiation layer is assumed to vanish when its temperature reaches the melting point of the material.

The model was applied to several case studies addressing different types of MLI (i.e., both polyester-based and aluminum-based ones) and fire scenarios to assess the effect on the heating rate and pressure build-up of the tank lading. The analysis allowed for the performance comparison of the studied MLIs, providing valuable information to support the emergency management of accidental scenarios involving liquid-hydrogen cryogenic tanks. Moreover, the results obtained can be used to define mitigation measures to protect the integrity of cryogenic tanks equipped with MLI.

30d What You Need to Know about Blue Hydrogen: Technology, Cost, and Safety

Anas Aqil, Sunhwa Park

Blue hydrogen emerges as an environmentally considerate energy production pathway, positioned centrally in the ongoing energy transition. Aimed at minimizing global CO_2 emissions, blue hydrogen is gaining momentum as a pivotal contributor to a cleaner energy paradigm. Its ascendancy is marked by innovation and practicality, establishing it as a promising contender in the move towards sustainable energy sources.

This review navigates through a multi-dimensional analysis of blue hydrogen, exploring technological advancements, economic considerations, and safety precautions. A comprehensive examination of prevailing technologies underscores the evolution and refinement observed in blue hydrogen production methodologies. Essential economic considerations are elucidated, offering insights into blue hydrogen's cost-effectiveness and affordability relative to alternative energy sources.

This study's central focus is the safety considerations associated with blue hydrogen practices. An exhaustive examination is conducted on the inherent risks and the corresponding safety measures currently implemented, identifying areas requiring further research and improvement in safety protocols.

In conclusion, this review aims to serve as a thorough resource, elucidating blue hydrogen's role, technological advancements, economic viability, and safety considerations in a comprehensive and accessible manner. The ultimate objective is establishing a well-rounded exploration that enhances understanding and knowledge in pursuing sustainable and safe energy solutions.

Recent Developments in Hydrogen Safety, Are We Headed to a Safer Fuel Alternative?

Kumar Israni

We continue to live in a world, where change is constant and challenges such a climate changes and world economy are constantly affected by the way industries operate. In the recent times, there need for alternate fuel sources has gained more momentum that ever. For the past few years, there have been tremendous discussions amongst learned groups on the use of hydrogen for this purpose.

While from an environmental perspective, this solution seems to a right choice, the safety side of things provide a very different perspective.

This paper will discuss the recent developments towards the safer use of hydrogen, and any regulatory changes that will impact the balance between the environmental and safety preferences over the use of hydrogen, using a risk based approach.

49a Safely Implementing an Explosive Hydrogen Market: Applying Lessons Learned to Expanding Hydrogen Usage

Achim Wechsung, Taylor Sulmonetti, Timothy Myers, Harri Kytomaa

With the DOE's announcement of the hydrogen hubs and the upcoming publication of the IRS guidance for the production tax credit for hydrogen, public and private investment in this sector is expected to grow. While hydrogen has a long and safe track record at large industrial hydrogen producers and consumers such as refineries, fertilizer plants, chemical facilities, and metal processing, hydrogen's unique physical properties require constant vigilance. Not only can unwanted hydrogen releases be easily ignited due to hydrogen's low minimum ignition energy, but hydrogen flames also travel more rapidly than those of other common flammable gases. As a result, hydrogen deflagration or detonation can result in more severe damage.

In this presentation, we will discuss strategies to prevent the release, ignition or explosion of hydrogen. Using two case studies, we will highlight critical handling steps during transport and/or usage. For example, when transferring hydrogen from a tube trailer, following proper isolation procedures is essential to prevent undesirable releases and potential ignitions. Alternatively, when charging a closed vessel such as a reactor with hydrogen, it is important to first purge air from the vessel. Each of these examples provides unique lessons learned that should be implemented or considered in future design and/or usage of hydrogen systems.

49b Addressing the Unique Challenges of Hydrogen Gas Detection and Its Way Forward in 3D Fire and Gas Mapping

Khama Matiti

Fixed Fire and Gas Detection Systems play a crucial role in enhancing safety at process plants, mitigating risks to personnel, the environment, and the facility. These systems are vital for detecting leaks, toxic or flammable gas build-ups, and the presence of fires, which could lead to hazardous situations. To optimize the placement of these detection systems, Fire and Gas Mapping software tools are increasingly employed.

These tools rely on specific performance criteria to analyze and optimize the placement of fire and gas detectors. For flammable gas detection mapping, one essential criterion is the size of the gas cloud or leak that needs detection. If the target gas cloud is too small, it may result in impractical, overdesigned systems that burden operators with maintenance. Conversely, if it's too large, the system may fail to detect critical containment breaches. Best practices and international guidance, such as the International Society of Automation (ISA) technical report (TR 84.00.07), offer methods for setting these criteria for hydrocarbon gas risks, but guidance for hydrogen gas risks is lacking.

Considering the growing interest in cleaner, zero-carbon hydrogen alternatives, there is a need to address the associated risks in 3D fire and gas mapping studies. Existing standards and guidance predominantly focus on hydrocarbon detection and do not adequately cover hydrogen-specific risks, such as its wider flammability range and higher ignition probability. This paper employs 3D computational fluid dynamics (CFD) explosion analysis to propose an effective strategy for defining appropriate performance targets for detecting flammable hydrogen gas. It also provides a case study to demonstrate how these targets can be used to establish a realistic and effective hydrogen gas detection system, reducing risks to As Low As Reasonably Practicable (ALARP).

49c Hydrogen Ignition Probabilities

Hai Thuy Minh Le

Hydrogen is expected to play a key role in the transition to a more sustainable low carbon energy system. As technologies are being developed in producing and handling hydrogen, ensuring safety is critical due to hydrogen's unique properties compared to those of traditional fuels. Hydrogen has a wide flammability range and low ignition probability which causes an increased risk of fire and explosion, particularly the transition from a deflagration to detonation upon ignition due to its high burning velocity. Therefore, identifying the probabilities of ignition of hydrogen and subsequence consequences upon ignition is a key factor in the risk assessment involving hydrogen handling and operation. This presentation discusses hydrogen ignition probabilities and proposal for the path forward.

A Study on Review of Safety Acts for Power Plants Using Hydrogen or Ammonia and Suggestion for Improvement Directions through Expert Survey

Da Hee Kim, Incheon National University Graduate School of Engineering, Incheon, Republic of Korea.
Hee Kyung Park, Incheon National University, Incheon, Republic of Korea.
Min Chul Lee, Incheon National University, Incheon, Republic of Korea.

*Email: ekgml92@inu.ac.kr (Presenting author), *LMC@inu.ac.kr (Corresponding author)*

Abstract

Hydrogen and ammonia are regarded as clean energy sources for realizing a decarbonized society with their versatile and potential. However, to use new fuels has been accompanied by various accidents. This not only affects human life, but also poses potential risks to the environment, so it is important to pay special attention to safety management when using new fuels. Therefore, this study examined safety-related laws for hydrogen and ammonia power plants to identify institutional aspects for applying new fuels to power plants. In addition, based on the analysis of the limitations of the current laws, a survey of 50 experts was conducted to suggest improvements to the legal system. Based on this study, it is to be expected that more studies on safety policies will be conducted to improve the safety of facilities that use new fuels and to prevent serious accidents.

1. Introduction

In order to realize carbon neutrality, the world is pursuing the conversion of existing carbon fuels to carbon-free fuels such as hydrogen and ammonia. In South Korea, the government, industry, and academia are rapidly advancing related research. In particular, various demonstration projects are underway to apply mixed LNG fuel with hydrogen and ammonia to gas turbine in combined cycle power plant(CCPP) that was previously operated with LNG. Carbon dioxide emission can be reduced if gas turbine in CCPP runs on a mixture of LNG with hydrogen and ammonia, which are carbon-free fuels [1-3].

However, despite the rapid introduction of decarbonized fuels into various industries under the lead of government, the appropriate legal system for the safety management of new fuels has not kept pace with the societal need. South Korea was the first country in the world to enact a law on the use of hydrogen called 「Hydrogen Economy Promotion and Hydrogen Safety Management Act」 (Hydrogen Act), but as the scope of hydrogen business continually expands, it is becoming clear that the Hydrogen Act cannot cover entire range. In addition, the inefficiency of Hydrogen Act is becoming an issue among experts [4]. Moreover, ammonia, which has been used in agricultural fields for more than 80% of its total volume [5, 6], has only recently begun to get attention as a decarbonized fuel [7, 8]. In other words, the applicable laws are vague or absent because there are no previous cases of its use as an eco-friendly energy. This means that new laws or improvements to existing laws are needed to reflect changes in usage.

If there are any deficiencies in the legal system, the demonstration period may be delayed or the project may be canceled, so in this study we attempted to review the legal system in advance and suggest direction to improve it. We reviewed the laws and regulations that apply to hydrogen and ammonia CCPP and analyzed their limitations. In addition, we utilized an expert survey to suggest solutions to improve the limitations found. We hope that this study will not only help to establish an institutional foundation for safer use of new fuels in achieving carbon neutrality, but also serve as a basis for establishing basic safety management system standards for facilities that use new fuels.

2. Research background

Hydrogen and ammonia plants are subject to the following laws: 「High-pressure Gas Safety Control Act」 (HGSC Act), 「Chemicals Control Act」 (CC Act), 「Occupational Safety and Health Act」 (OSH Act) and 「Hydrogen Economy Promotion and Hydrogen Safety Management Act」 (Hydrogen Act). After reviewing the criteria and details of each law, [9] we found four limitations.

First, the Hydrogen Act is for low-pressure hydrogen (compressed gas < 1 MPa, liquid gas < 0.2 MPa), so it does not apply to hydrogen CCPP using high-pressure hydrogen (compressed gas ≥ 1 MPa, liquid gas ≥ 0.2 MPa). Second, gas turbines in hydrogen and ammonia CCPPs handle large volumes of fuel, but, according to the HGSC Act, hydrogen and ammonia CCPP is not subject to evaluate safety and submit safety improvement plan. Third, as shown in **Table 1**, the HGSC Act has four different storage and processing capacity standards for facilities classified as general production, but only two different standards for the appointment of safety managers for such facilities. In other words, it can be interpreted that the standards for safety are eased. Finally, as shown in **Table 1**, the criteria for the appointment of a safety manager under the HGSC Act allows appointment of safety staff who have completed the general facility safety staff training course. Verification of the appropriateness of these qualifications for the facilities using new fuels is required.

Table 1. Standards of safety manager appointment for general production facility

Storage and processing capacity	Appointment of safety manager	
	Kind and number	Qualifications
·Storage capacity > 500 ton	1 safety officer	Industrial engineer gas
·Processing capacity > 2,400 Nm^3/h	2 safety staffs	Complete training program
·100 ton < storage capacity ≤ 500 ton	1 safety officer	Industrial engineer gas
·480 Nm^3/h < processing capacity ≤ 2,400 Nm^3/h	2 safety staffs	Complete training program
·100 ton ≤ storage capacity	1 safety officer	Craftsman gas
·60 Nm^3 < processing capacity ≤ 480 Nm^3	1 safety staff	Complete training program
·Processing capacity ≤ 60 Nm^3	1 safety officer	Craftsman gas
	1 safety staff	Complete training program

3. Method: expert survey

Based on the limitations analyzed in Section 2, we developed a questionnaire to seek improvement directions for the law and surveyed 50 experts. The questionnaire was designed with a total of 13 multiple-choice questions, including 2 questions on the risk perception level of hydrogen and ammonia fuels, as well as 11 questions on the adequacy of existing laws applied to new fuel CCPPs and the need for improvement. To derive professional improvement directions, a survey was conducted among experts engaged in R&D, safety management, fuel production and operation, plant design, fuel use, etc. at universities, research institutes, and manufacturing companies related to hydrogen and ammonia research and industry. The demographics of experts who participated in the survey are shown in **Fig. 1**.

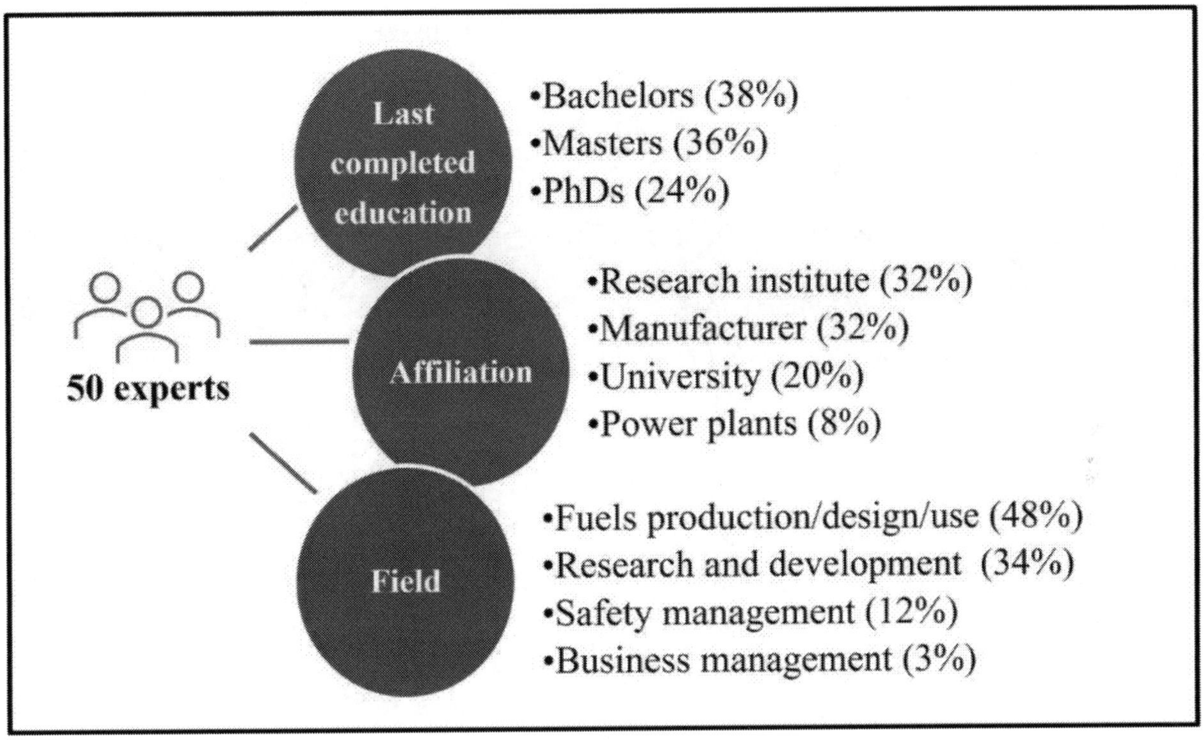

Fig. 1. The demographics of the experts who participated in the survey

4. Results

First, we examined experts' perceptions of the risk level of hydrogen and ammonia compared to conventional LNG fuels. We posed the question to experts: If LNG has a comprehensive risk score of 50, how much is the comprehensive risk score for hydrogen and ammonia? Experts ranked the risks in the following order: hydrogen (72) > ammonia (65) > LNG (50). Experts consider new fuels to be more dangerous than traditional fuels.

Next, regarding the adequacy of the existing laws and regulations governing new-fuel CCPPs and the need for improvements, most experts stated that the existing legal system is still inadequate and needs to be improved. To confirm the survey results quantitatively, we scaled the median to 50 and analyzed. Values above 50 indicate that the content needs to be improved, while values below 50 indicate that it is unnecessary. According to the survey results, the need for improvement was highest in the following order: enactment of the Ammonia Act (78) > designated as a comprehensive safety management subject under the HGSC Act (73) > strengthening of existing laws and technical codes = enactment of emergency training drills = classification of ammonia as a Class 4 hazardous substance under the 「Act on the Safety Control of Hazardous Substances」 (SCHS Act) (70) > review of the criteria for appointing safety managers under the HGSC Act and the Hydrogen Act (59). All values are above 50, clearly indicating the need to review and improve the current legal framework.

Most of all, the demand for the enactment of an Ammonia Act like the Hydrogen Act was the highest. Therefore, we reviewed the opinions of experts on the Hydrogen Act that can be used as a reference for the enactment of the Ammonia Act. Experts responded that the scope of Hydrogen Act is inadequate. Currently, CCPPs are divided into sections based on shut-off valves and different laws apply to each area. Experts pointed out that this coverage is inadequate and that all laws should be unified and applied throughout CCPPs. This means that the Hydrogen Act, which is the first law targeting new fuels, needs to be reviewed and improved prior to the enactment of the Ammonia Law, and efforts should be made to consolidate content scattered in other laws.

Based on the results of the above expert survey, we have identified four recommendations to improve the current limitations of the Korean legal system. First, scattered laws must be integrated into one. Second, the legal system must continuously evolve with technological advancements, and for ammonia fuel, new laws and regulations need to be revised. Third, Risk assessment and safety manager appointment standards must be improved. Fourth, safety laws should be expanded to scope of prevention and education.

5. Conclusion

Technology development for introducing new fuels into power plants is progressing rapidly with several demonstration projects. However, as shown in the expert survey, despite the high concern about the risks of new fuels, institutional research on infrastructure, policies, and laws has not caught up with the pace of technological development. In fact, the legal standards and content applied to power plants also reveal various limitations. Multifaceted efforts are needed to reduce carbon emissions and revitalize the hydrogen and ammonia economy by developing carbon-free power generation technologies that are on par with existing LNG power generation. In order to successfully commercialize the technology, it is necessary to focus not only on technology development but also on institutional research for safe infrastructure construction and safety management systems. The first hydrogen law in the world should be continuously supplemented with advances in technology, and existing laws should be reviewed and improved. In addition, enactment of laws should be considered when necessary. Moreover, as the risk perception is high, stronger safety standards are needed, and effective safety management measures through education and training are required.

6. Acknowledgement

This work was supported by Korea Institute of Energy Technology Evaluation and Planning(KETEP) grant funded by the Korea government(MOTIE) (RS-2023-00270080, Development of a method for testing and evaluating hydrogen turbine combustor, 50%), and this research was supported by Korea Electric Power Corporation (Grant number : R22XO02-06).

7. Reference

1. Noussan M, Raimondi PP, Scita R, Hafner M (2021), "The role of green and blue hydrogen in the energy transition—a technological and geopolitical perspective", Sustainability (Switzerland), pp. 1–26.
2. Wasiak A, Orynycz O, Tucki K, Świć A (2022), "Hydrogen Enriched Hydrocarbons as New Energy Resources – as Studied by Means of Computer Simulations", Advances in Science and Technology Research Journal, pp. 78–85.
3. Hammerstrom B, Niezrecki C, Hellman K, Jin X, Ross MB, Mack JH, et al (2022), "The viability of implementing hydrogen in the Commonwealth of Massachusetts", Front Energy Res.
4. Lee Y, Lee MC, Kim YJ (2022), "Barriers and strategies of hydrogen fuel cell power generation based on expert survey in South Korea", Int J Hydrogen Energy, pp. 5709–19.
5. Joseph Sekhar S, Samuel MS, Glivin G, Le TG, Mathimani T (2024), "Production and utilization of green ammonia for decarbonizing the energy sector with a discrete focus on Sustainable Development Goals and environmental impact and technical hurdles", Fuel.
6. Cui P, Zhou Y, Song T, Xu Z, Zhang J, Liu Y, et al (2023), "Thermodynamic and Economic Analysis of an Ammonia Synthesis Process Integrating Liquified Natural Gas Cold Energy with Carbon Capture and Storage", ACS Sustain Chem Eng, pp. 3052–64.
7. Ishaq H, Crawford C (2024), "Review and evaluation of sustainable ammonia production, storage and utilization", Energy Convers Manag, pp. 117869.
8. Lin B, Nowrin FH, Rosenthal JJ, Bhown AS (2023), "Malmali M. Perspective on Intensification of Haber–Bosch to Enable Ammonia Production under Milder Conditions", ACS Sustain Chem Eng, pp. 9880–99.
9. Ministry of government legislation Korean law information center Retrieved from https://www.law.go.kr

AUTHOR INDEX

AIChE
120 Wall Street, FL 23
New York, NY 10005-4020

ISBN 979-8-3313-1713-3

CURRAN ASSOCIATES INC.
proceedings.com